Stainless Steels In Industry: Stress Corrosion Cracking of Austenitic and High Nitrogen Stainless Steels

Introduction

Stainless steels are ferrous alloys which contain a minimum of 10% chromium. The purpose of chromium is to provide a composition that usually forms a protective layer on the steel metal surface; in other words, it allows steel passivity. This property makes stainless steel one of the best alloys for corrosion and oxidation resistance. Many types of stainless steel also contain 8 - 10% nickel, which makes them flexible.

Stainless steels can be divided into five main groups: Austenitic, ferritic, duplex, martensitic, and precipitation hardening. Table I lists the compositions of some common type of stainless steels of each group. The Unified Numbering System (UNS) designations for stainless steels consist of the letter "S" followed by five numbers, which usually incorporate the numbers of the more common alternative designations. Thus S 30400 is the austenitic stainless steel commonly designated AISI type 304 and S 44600 it the ferritic steel AISI type 446.

Table I: Alloy Designations for Some Common Stainless Steels.

Type	UNS Number	Composition (wt.)[a]							
		C	Mn	Si	Cr	Ni	Mo	Cu	Al
Austenitic types									
201	S20100	0.15	5.5-7.5	1.00	16.0-18.0	3.5-5.5			
304	S30400	0.08	2.00	1.00	18.0-20.0	8.0-10.5			
310	S31000	0.25	2.00	1.50	24.0-26.0	19.0-22.0			
316	S31600	0.08	2.00	1.00	16.0-18.0	10.0-14.0	2.0-3.0		
347	S34700	0.08	2.00	1.00	17.0-19.0	9.0-13.0			
Ferritic types									
405	S40500	0.08	1.00	1.00	11.5-14.5				
430	S43000	0.12	1.00	1.00	16.0-18.0				
Martensitic types									
410	S41000	0.15	1.00	1.00	11.5-13.0				
501	S50100	0.10 min	1.00	1.00	4.0-6.0				
Precipitation-hardening types									
17-4 PH	S17400	0.07	1.00	1.00	15.5-17.5	3.0-5.0		3.0-5.0	
17-7 PH	S17700	0.09	1.00	1.00	16.0-18.0	6.5-7.75			0.75-1.5

Source: Data from *Metals Handbook*. 9th ed.; Vol. 3. American Society for Metals. Metals Park: Ohio. 1980.

[a] Single values are maximum values unless otherwise indicated.

Stainless Steel in the history

Stainless Steel was discovered by the english metallurgist Harry Brearley around 1913. Martensitic was the first stainless steel with 0.24% carbon and 12.8% chromium. Within a year of Brearley's discovery, Strauss & Maurer in Germany developed the first austenitic stainless-steel grades. Dansitzen, in the United States, discovered the ferritic stainless steel in 1915.

The first stainless steel duplex was produced for application in the paper industry in Swede, around 1930. The next stage in the development of stainless steel was in the late 1960s, where the Argon-Oxygen decarburisation process (AOD) made it possible to produce much cleaner steel with a low carbon level [1]

Today there are hundreds of stainless-steel grades available for countless applications. Their corrosion resistance, mechanical properties, and cost vary over a wide range. For this reason, it is important to specify the exact stainless steel desired for a given application.

Production of Stainless Steel

Stainless steel is produced in an electric arc furnace where carbon electrodes contact recycled stainless scrap and various alloys of chromium, nickel, molybdenum, etc. depending on the stainless type. Current is transmitted through the electrode and the temperature increases to a point where the scrap and alloys melt. The molten material from the electric furnace is then transferred into an AOD vessel, where the carbon levels are reduced (stainless steels have a much lower carbon level than mild steels) and the final alloy additions are made to make the exact chemistry.

The process also continues from melting and casting either into ingots or continually cast into a slab or billet form. Then the material is hot rolled or forged into its final form. Some material receives cold rolling to further reduce the thickness as in sheets or drawn into smaller diameters as in rods and wire. Most stainless steels receive final annealing (a heat treatment that softens the structure) and pickling (an acid wash that removes furnace scale from annealing and helps promote the passive surface film that naturally occurs) [2].

Passivity of Stainless Steel

The presence of a 'passive' oxide film on the surface makes stainless steels good corrosion-resistant alloys. Although, this film is extremely thin 1-5 nm (1-5 $\times 10^{-6}$ mm) and invisible, it is adhered firmly, self-healing, and is chemically stable. The passivity of metals and alloys has been a subject of investigation for many years. The polarization curve (Fig. 1.0) presents a clear understanding of the passivity of theory.

The electronic device potentiostat, Gamry Potentiostat-Reference 600 (Fig. 2.0), which controls potential, gives measurements of the current density and potentials required for plotting polarization curve. To get passivity, the potential of the electrode should be raised from E_{corr} (corrosion potential) to E_{pp} (primary passivation potential). Above E_{pp}, the current drops suddenly to a low value, i_{pass}

(the passive current density), and remains at low value over a wide range of potentials. This range is termed the potential passivity range. The fall in current density can be attributed to the existence of a thin passive film, which has high electrical resistance. This passive film usually has a thickness range from 1 to 10 nm for all metals. Therefore, corrosion rates decrease to low values. Passivity film has some electric conductivity. They are not insulators; consequently, only small potentials can be maintained across passive film (approximately 1 volt).

On continuing to increase the applied potential, another potential will be reached, above E_p (pitting potential), at which the measured current will again begin to increase. As a result, the passivity breakdown and then severe corrosion may occur. The chromium content in stainless steel provides a composition, which will typically develop a protective layer on the surface of the steel metal; in other words, it allows passivity to the steel.

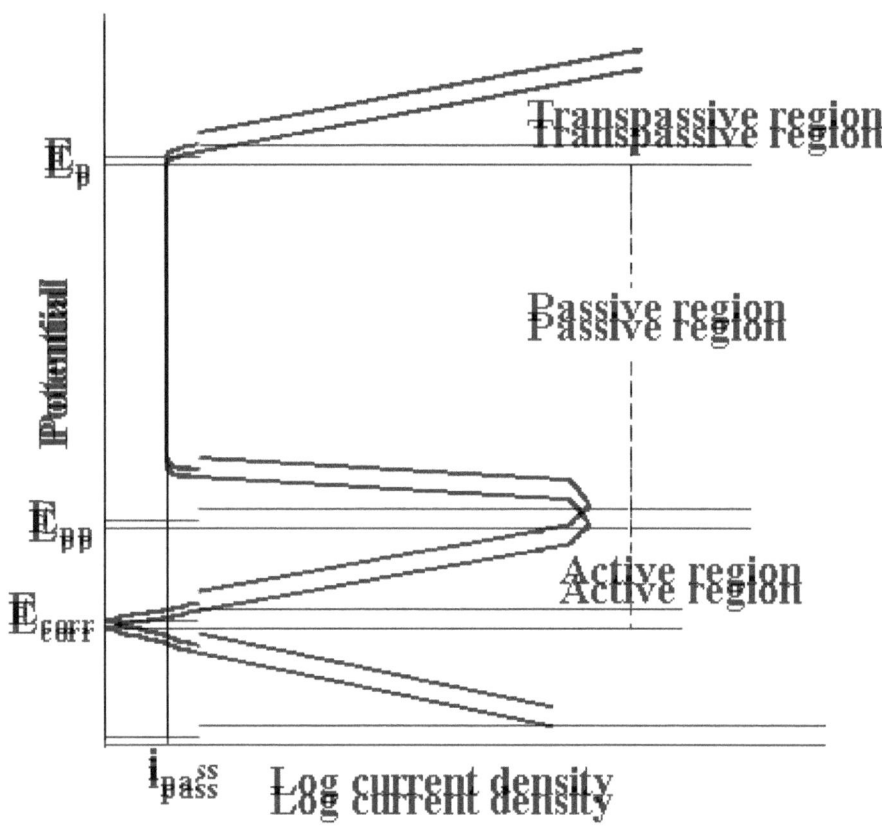

Fig. 1.0: Polarization Curve Exhibiting Passivity.

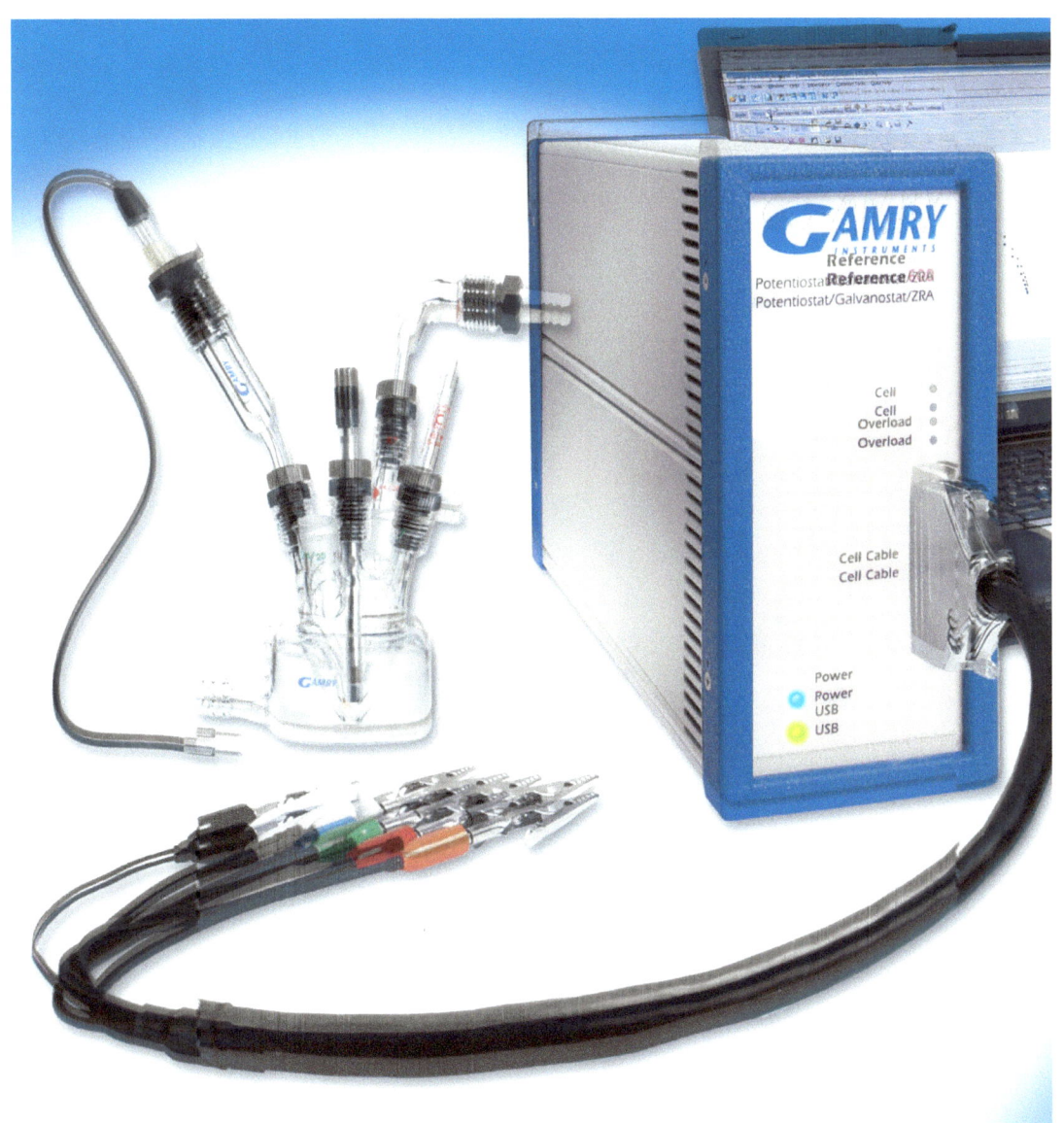

Fig. 2.0: Gamry Potentiostat-Reference 600.

Austenitic Stainless Steels

Austenitic stainless steels are iron-chromium-nickel alloys. These alloys contain 18-25 wt% Cr and 8-20 wt% Ni and low C. They may also have additions of Mo, Nb, or Ti. Austenitic stainless steels are non-magnetic and cannot be hardened by heat treatment. They can be hardened only by cold working and show an increase in strength as a result of cold work. Austenitic alloys are excellent formability and are readily in fabrication, including welding.

Austenitic stainless steels have a good combination of mechanical strength, fabricability, and excellent corrosion resistance. Austenitic alloys constitute up to 70% of the total stainless steels in use. They are extensively used as construction material in the chemical, petrochemical, fertilizer industry. Austenitic steels are widely specified for the more severe corrosion conditions such as those encountered in handling nitric acid and nuclear industries.

Type 304 is the most frequently using among various kinds of austenitic. Its nominal composition is 18 Cr-8 Ni; therefore, it is often referred to as 18 -8 Stainless Steel. This type has good atmospheric corrosion resistance, and hence it's widely used in the food and beverage industries and for medical purposes. It is also used extensively for chemical processing equipment [3].

Type 316 is alloyed with 2% to 3% molybdenum to improve pitting and crevice corrosion resistance compared with 304. It is extensively used for food, chemical processing, pulp, and paper equipment. Type 316 is one of the most candidates for marine applications.

One of the primary defects in austenitic stainless-steel applications is its susceptibility to chloride CSCC, especially. Within a few months, the stressed austenitic may crack easily at a relatively high temperature above 60 C in an environment that contain few ppm of chloride ions (and oxygen oxidizers) [4]. This problem frequently occurs in petroleum refineries, chemical plants, food processing industries, and marine environments. Instead, using ferritic stainless steel is one of the best solutions to this problem.

SCC can occur easily under a specific condition in the laboratory, for example:

Type 304 +applied stress 50 Kpsi in boiling MgCl (within 30 minutes only SCC occurs).

Type 316 + applied stress 50 Kpsi in boiling Mg Cl (within 2 hrs SCC occurs).

Austenitic stainless steels have been used successfully in many applications in marine environment. Types 304 and 316L are the most likely candidates for marine applications due to their excellent corrosion resistance especially for pitting and crevice. Therefore, they are found in excess in work-boat propellers, pump components, valves, shaft components, hull fittings, fasteners and oceanographic instruments [5]. Type 316L stainless steels are considered to be the main components in hydraulic control systems for the operation of subsea oil recovery system. These alloys also are used in instrument and chemical injection tubing in offshore oil platform [6;7].

Ferritic Stainless Steels

Ferritic Stainless Steels are iron-chromium alloys with 15-30 wt% Cr, low C, no nickel. Ferritic stainless steels are inherently magnetic. These steels, like austenitic steels, cannot be hardened by heat treatment; they are hardenable only by cold working. Ferritic stainless steels have good ductility and can be formed readily.

Ferritic stainless steels are highly resistant to chloride stress-corrosion cracking. Type 430 with about 17% Cr, has excellent atmospheric corrosion resistance. Therefore, it is used extensively for automotive trim and cooking utensils. They are also used for other applications in which welding is not a requirement.

Types 442 and 446 are highly Chromium content (18-27% Cr) which make them good resistant to high-temperature oxidation and sulfur gas attack. For this reason, they are found frequently in furnace parts and heat-treating equipment. The new material-type 444 (18 Cr- 2 Mo) has good resistance to pitting and crevice corrosion. It is equivalent to type 316 in many environments.

The sensitized ferritic stainless steel, like austenitic, is susceptible to intergranular stress corrosion cracking. Using low carbon grades or ordinary steel may help in reducing this problem. $MgCl_2$ solution usually causes transgranular SCC.

Duplex Stainless Steels

The duplex stainless steel contains austenite and ferrite in equal proportions. The balance between ferrite and austenite is achieved by adjusting the amounts of Cr (18-26 wt.%), Ni (5 - 6 wt.%), Mo (1.5-4 wt.%) and nitrogen. Duplex stainless steels have significantly higher yield strength than the austenitic grades and have better toughness than do the ferritic.

Because the duplexes have better toughness than do the ferritic, they are available in plate thickness and may, therefore, be used for the tube sheets. Their resistance to chloride stress-corrosion-cracking is in between that for the ferritic and austenitic and decreases with increasing cold working.

The crevice corrosion resistance of these steels is somewhat less than that of ferritic or austenitic grades of equivalent chromium and molybdenum compositions. Duplex stainless steels are often used for marine applications such as chemical tankers, offshore oil production facilities and fasteners.

Martensitic Stainless Steels

Martensitic stainless steels are iron-chromium-carbon alloys. These alloys generally contain 11-15 wt.% Cr and 0.1-1.2 wt. C%. They can be hardened like ordinary carbon steel, by heat treatment. Strength increases and the ductility decreases with increasing hardness. They have high strength with low toughness. Martensitic stainless steels are magnetic and have good ductility.

The martensitics are used in applications requiring moderate corrosion resistance plus high strength or hardness. They are used for joints such as turbine blades, aircraft fittings, bolts, pump shafts, valves and bearings. They are also used for cutlery, knives, razor blades, springs and surgical instruments.

The martensitic are extensively used for the mildly corrosive chemicals and wet or dry corrosion environments found in steam. Type 410 is the most widely produced grade and is considered a general-purpose martensitic. Type 416 is more comfortable to cut and is used for valve steams, nuts, bolts and other parts to reduce machining costs.

Precipitation-Hardening Stainless Steels

Precipitation-hardening stainless steels are chromium-nickel grades containing other alloying elements such as copper or aluminum. Precipitation-hardening stainless steels are magnetic. They can be hardened to high strengths by solution treatment and aging at a moderately elevated temperature.

The precipitation-hardened grades must not be subjected to further exposure-elevated temperature by welding or during service, because averaging of the precipitation can result in loss strengthening. These steels are used for gears, fasteners, cutlery and aircraft parts. Typical examples of this class of stainless steels are 17-7PH, PH 15-7Mo and AM-350.

Sensitization in Stainless Steels

When stainless steels are extensively heated or slowly cooled through the temperature range of 450 °C to 850 °C, chromium carbide (Cr_3C) precipitates along the grain boundaries. This leads to subsequent chromium depletion in the regions adjacent to the grain boundaries (Fig. 3.0). As a result, the protection afforded to the alloy by the presence of chromium is lost, and severe intergranular corrosion occurs. Sensitized Austenitic usually exhibits Intergranular whereas un-sensitized Austenitic often exhibits transgranular of CLSC [8].

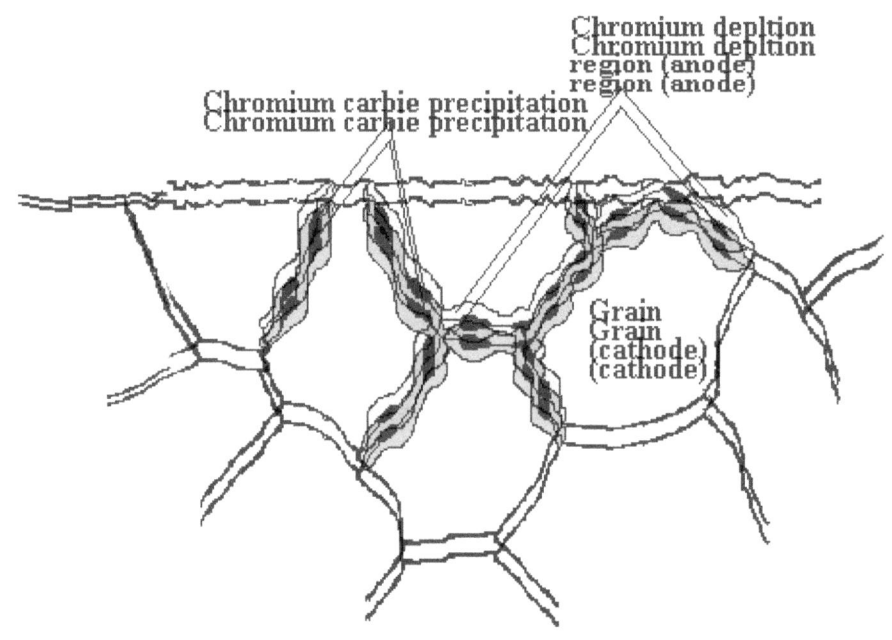

Fig.3.0: Intergranular corrosion caused by precipitation of chromium carbide in the grain boundaries

The most critical temperature is about 650 °C; where holding for only a few seconds may be sufficient to permit subsequent deterioration. This can easily happen during the welding process.

Sensitization is common in austenitic stainless steels. It also occurs in ferritic stainless steel with a less degree. This problem frequently occurs in petroleum refineries, chemical plants, food processing industry and marine environment. Sensitization and intergranular corrosion usually extent to the adjacent area of the weldment (heated affected zone HAZ). This phenomenon is commonly called "weld decay" (Fig: 4.0):

Fig. 4.0: Weld Decay (304SS)

Solution of sensitization problem:

The following methods can avoid sensitization of austenitic stainless steel:

1. Austenitic stainless steel may be heat treated (annealing): Heat to 1065 °C followed by quenching in water or rapid cooling to room temperature.
2. Using low carbon grades such as L304 and L316.
3. Using developed ferritic stainless steel is one of the best solutions to this problem

4. Using strong carbide former such as titanium Ti, columbium Co or tantalum Ta. These elements have a strong tendency to form carbides.

Therefore, when the steel is heated, carbon will combine rapidly with these elements not with chromium. Thus, the chance for chromium depletion will eliminate.

In case of Ferritic stainless steel, there are several solutions:

1. Heating the steel to 790 °C for an hour or so followed by quenching to room temperature.
2. Annealing at 840 °C few hours to homogenize the chromium.
3. Using high chromium alloys such as 321, 347.
4. Using strong carbide former such as Ti, Co, Ta.

New Development In Stainless Steels Alloys

New grades of stainless steels with higher alloy contents have been designed to meet the demand for higher corrosion resistance in special applications. Theses alloys are Super Stainless-Steel Alloys and High Nitrogen Stainless Steel.

Super-Austenitic Stainless Steel

These alloys were designed to have superior stress corrosion cracking and pitting resistance relative to the standard austenitic stainless steels. This achieved by increased levels of nickel, molybdenum and nitrogen.

Most of the super-austenitic grades lie within the following composition limits: 20-25 Cr, 15-25 Ni, 4-8 Mo, 0.01-0.03 C, and 0.2 –0.6 N.

Super austenitic grades of stainless steel contain high levels of molybdenum 5% to 7% (some times are referred to as 6% molybdenum or 6 moly). Many of these alloys were developed for severe marine applications, such as electric power plant, condensers and oil production facilities in the North Sea. The 6% molybdenum superaustenitic stainless steels are much more resistant to chloride SCC than the ordinary 304/316 types. Also, these alloys are resistant to corrosion under highly corrosive conditions, including seawater immersion at ambient and elevated temperature [9]. However from an economic standpoint molybdenum is very expensive and the market for stainless steels is very competitive. Therefore, selection of stainless steels with higher content of nitrogen for severe marine applications, such as power generation plants, heater tubes and offshore oil production facilities seems wiser. This will be descused in section of effect of Nitrogen on austenitic stainless steel.

Super-Ferritic Stainless Steels

Super-ferritic grades have excellent marine corrosion resistance and are used in tubing for seawater condensers at power plants.

Super-Duplex Stainless Steels

Super-duplex stainless steels are highly corrosion resistant to crevice corrosion. These alloys contain 25-27 Cr; 5-7 Ni; 4-6 Mo; 0.01-0.03 C, and 0.15-0.25 N. Some alloys also contain tungsten and copper. Super-duplex stainless steels have been widely applied in the oil, gas and petrochemical industries.

Super-Martensitic Stainless Steels

These alloys contain 11-13 Cr; 4-6 Ni; 1-3 Mo and 0.1-0.03 C. The low carbon content improves weldability by ensuring resistance against hydrogen-induced cracking. These alloys have been used for oil and gas pipeline applications. The combination of strength and corrosion resistance makes them potential replacement alloys for duplex stainless steels.

High Nitrogen Stainless Steel (HNSS)

High Nitrogen Stainless Steels (HNSS) belong to the category of steels and stainless steels containing nitrogen. HNSS alloys have become an important class of engineering materials and have received much attention in recent years. The effects of nitrogen on the properties of steels have long been a subject of study. Theses alloys have outstanding mechanical properties such as strength, toughness, creep resistance and non-ferromagnetic behavior. HNSS has excellent resistance for general corrosion and SCC.

Nitrogen alloyed steels are considered HNSS when the nitrogen content exceeds 0.4 wt% in the austenite matrix and 0.08 wt% in the case of ferritic structure. Nitrogen alloying in stainless steels is known to have many beneficial effects, including improvements in phase stability, strengthening and corrosion resistance [10].

These alloys have been found to have increasing applications in:

1= Power generation industry and energy production
2= Chemical and petrochemical industries including many chemical equipment's
3= Petroleum and nuclear industries such as in Oil and gas refineries
4= Marine sectors and ship building
5= Mining sectors
6= Transportation sectors

Nitrogen content up to 0.16 wt.% is reported to improve sensitization resistance. It was also reported that in 316 SS alloys, as nitrogen content increases, the time required for sensitization increases from 0.5hr as much as 80 hr [10].

Effect of Nitrogen Content on Stress Corrosion Cracking of Austenitic Stainless Steels in Seawater

In many marine applications, the corrosion resistance of the traditional Types 304/316L stainless steels may not be sufficient, particularly for more demanding applications such as continuous immersion or exposure at elevated temperatures. It has been shown by the Welding Research Council [11] that 300 series stainless steels, heavily sensitized by furnace heat treatment, displayed intergranular corrosion in ambient sea-water exposures. Austenitic stainless steels are generally susceptible to stress corrosion cracking (SCC) in chloride containing environments at high temperature. This type of cracking, defined as chloride stress corrosion cracking (CSCC), is mostly transgranular in nature and is not affected by the change in metallurgical structure of alloys [12].

The role of nitrogen in the passivity of stainless steels has been explained through several mechanisms. Nitrogen alloying has been observed to retard localized corrosion initiation and to suppress the growth of the localized corrosion attack effectively

It is well known that, alloying of nitrogen in stainless steels improves a number of corrosion properties. However, the mechanistic role of nitrogen in improving the corrosion resistance is still under intensive investigation and need more study. Nitrogen alloying improves the cavitation erosion resistance of austenitic stainless steels in ocean. The main reason for the increase in cavitation erosion resistance of austenitic stainless steels is the effect of nitrogen on their hardness [13].

An attempt was made by Almubarak *etal.* [14] to investigate corrosion behavior for different grades of stressed sensitized austenitic stainless steel in seawater. Types 304 and 316L stainless steels are traditional types and commonly used in marine environment applications. These types have been compared with types 304NH and 316NH austenitic stainless steels to study the effect of high nitrogen content on the corrosion resistance.

Immersed specimens in seawater were subjected to a constant tensile stress. Experimental work that included corrosion rate, microstructure investigation and electrochemical test has been carried out for the stressed sensitized austenitic stainless steel specimens. Seawater was taken from the deep sea of Arabian Gulf to be used for that research.

Here we try to demonstrate a brief summary of the experimental work of Almubarak *etal.* team for the corrosion behavior of the austenitic stainless steels of high nitrogen content in Seawater [14]. A series of tests was carried out to evaluate five types of austenitic stainless steels in seawater. The materials studied were commercial austenitic stainless steels types: 304, 316L, 304NH and 316NH. Types 304NH with Nitrogen content 0.52% weight and 316NH with Nitrogen content 0.64% weight, are austenitic stainless steels alloys which have high nitrogen content, have been selected to study the effect of nitrogen on the corrosion behavior. Stainless steel specimens are subjected to temperature of 550 °C, such a temperature with long time exposure, is enough to create sensitization inside the grain boundaries.

Apparatus has been designed and installed especially for this research. The apparatus has a beam of rectangular shape supported by two horizontal beams. A pull cylinder (Enerpac BRP-106C) has been hanged from the upper horizontal beam by a hook. A tube was connected between the pull cylinder and a hydraulic hand pump to exert a pressure inside the cylinder. The immersed specimen in seawater was subjected to a constant tensile stress of 240 MPa. A stainless steels bucket with bubbling oxygen inside the seawater was fixed and hold by two bending steel bars from both sides. The top-side end of specimen was connected, by a longer chain, to the pull cylinder as shown in Fig. 5.0.

The specimen was subjected to electrochemical potentiokinetic reactivation (EPR) test and an anodic polarization curve was recorded. The specimen was then subjected to microstructure investigation to determine the type of corrosion

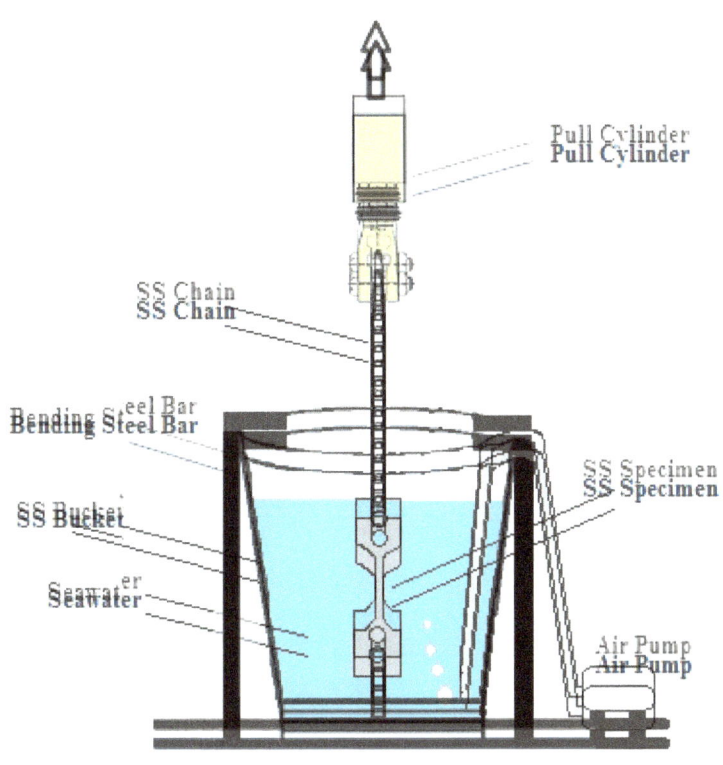

Fig. 5: Drawing shows the position of the specimen in the steel bucket.

In that research (Almubarak et.al. 14) investigation, EPR tests have been applied to establish anodic polarization curves for stainless steel specimens. The electrochemical test equipment consisted of a potentiostat (GAMRY 300), a stainless steel specimen as a working electrode, a platinum counter electrode 1.50 in^2 area and a reference electrode (saturated calomel electrode, SCE).

Discussion and Results

Microstructure investigations for the stressed sensitized type 304 in seawater is shown in Fig. 6-a. A severe chloride stress corrosion cracking CSCC has been seen noticed in the specimen. The microstructure investigation showed the presence of branching intergranular and transgranular cracks. The microstructure of the specimen type 316L showed a classical branching CSCC, where the transgranular nature was obvious (Fig. 6-b). However, intergranular stress corrosion cracking was also noticed in the specimen structure.

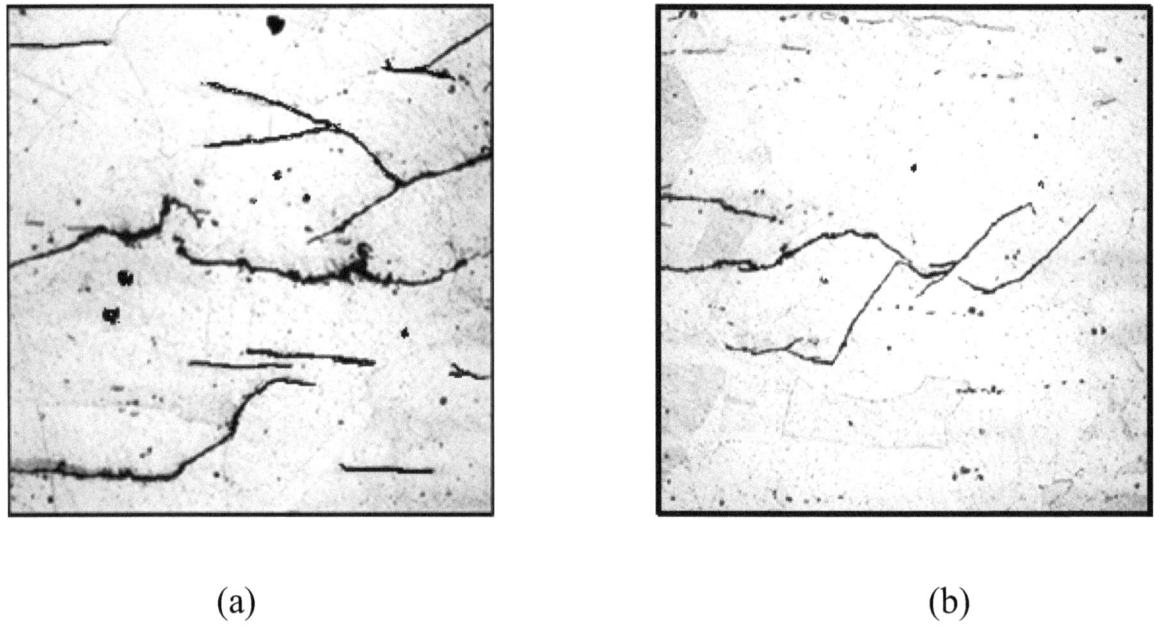

(a) (b)

Fig. 6: (a) Austenitic stainless steel type 304 (80 µm), showing a sever chloride stress corrosion cracking and several large pits.
(b) Austenitic stainless steel type 316L (80 µm), showing a classical chloride stress corrosion cracking and small size pits.

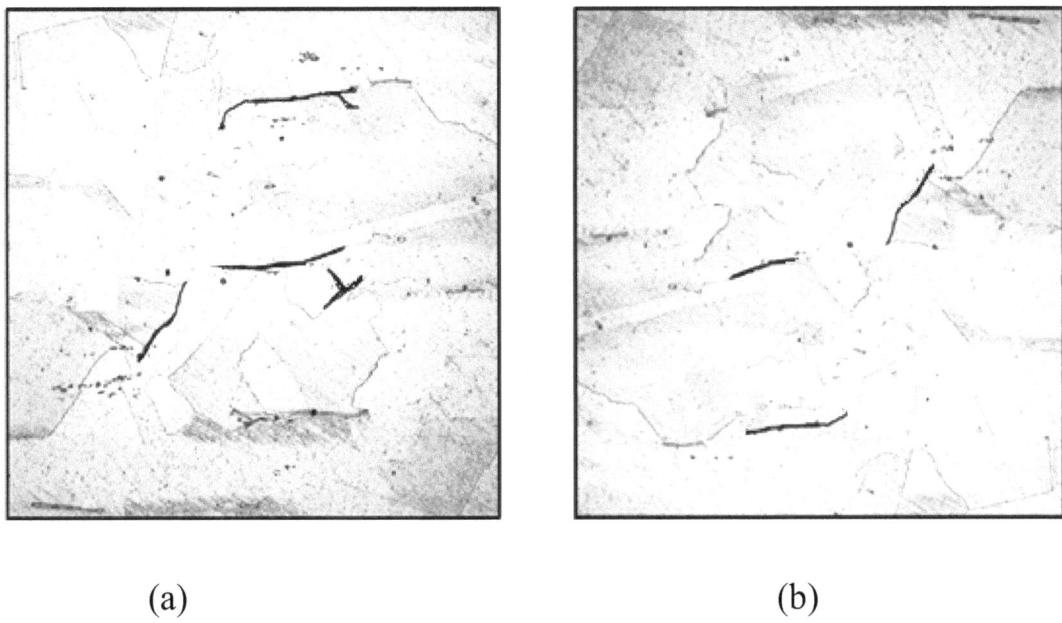

Fig. 7: (a) Austenitic stainless steel type 304NH (80 μm), showing a very slight stress corrosion cracking and few small pits.
(b) Austenitic stainless steel type 316NH (80 μm), showing hair line cracks and no pits.

The beneficial effect of alloying nitrogen in austenitic stainless steels appeared clearly in the two types 304NH and 316NH. Only very slight SCC has been observed in the microstructure examination (Figure 7-a) for the austenitic type 304NH. Whereas, hair line cracks appeared in Fig. 7-b for the type 316NH. Microstructure investigation for the sensitized specimens led to a result that austenitic stainless steels of high nitrogen content are excellent in resisting stress corrosion cracking in seawater.

Pitting has been noticed also, through microstructure investigations for the stressed sensitized specimens in seawater. Fig. 6-a showed several significant pits in type 304. Fig. 6-b showed some small size pits in type 316L. Again, the effect of high nitrogen content in stainless steels can be noticed clearly in types 304NH and 316NH. Fig. 7-a showed few small pits whereas Fig. 7-b showed almost no pits for the microstructures 304NH and 316NH respectively.

The pitting resistance equivalent (PRE) of stainless steels containing nitrogen is defined as PRE= 1 [Cr] + 3.3 [Mo] + x [N]; with x being between about 13 and 30 [15]. It has been reported that using a coefficient of 30 for nitrogen allowed the best fit to experimental data for a wide range of nitrogen-alloyed austenitic steels [16]. Table IV showed high PRE values for types 304NH and 316NH due to their higher nitrogen content. Microstructure investigation results agreed well with the correlation of PRE; which attributes that stainless steels of high nitrogen content increases the resistance to pitting.

The corrosion rate was calculated for each specimen according to the following formula:

$$\text{Corrosion Rate (mpy)} = \frac{534 \times W}{DAT}$$

Where, W, is weight loss in (mg); D, is density of specimen in (g/cm³), A, is total area of specimen for both sides in (in²) and, T, is time exposed in (hrs). Table II shows corrosion rate for the specimens after 120 days in seawater.

As it was expected, corrosion rate values were almost negligible and limited between the different grades of austenitic stainless steels. However, the estimated values indicated to reasonable remarks. Corrosion rate was less in type 316L than that of type 304 due to its higher molybdenum content. Obviously, the specimens 304NH and 316NH reported less corrosion rate values than the other grades. This behavior assures that high nitrogen content assist to create resistance to the corrosion process in the alloy steel.

Table II: Corrosion rate for specimens in seawater in (mpy) and pitting resistance equivalent (PRE)

Type of SS	304	316L	304NH	316NH
Corrosion Rate	2.460	1.681	0.661	0.428
PRE	19.18	24.96	34.80	40.76

It has been concluded that stressed sensitized austenitic stainless steels of high nitrogen content, corroded at a much slower rate, increased the pitting resistance and exhibited an excellent resistance to stress corrosion cracking in seawater.

The general experience in marine environment with austenitic stainless steel equipment's has showed a successful operation in sweater at moderate's temperatures (i.e. < 100 ºC). Austenitic stainless steel of high nitrogen content can be more satisfactory for marine sectors and offshore service such as heater tubes, shipbuilding and power generation in marine tankers where equipment's are exposed to tensile stresses and elevated temperatures.

Stress Corrosion Cracking of Austenitic Stainless Steels In Petroleum Refineries

As mentioned previously Austenitic alloys constitute up to 70% of the total stainless steels in use. Austenitic stainless steels are generally susceptible to SCC in chloride containing environments often defined as chloride stress corrosion cracking (CSCC). This type of cracking is mostly transgranular in nature and is not affected by the change in metallurgical structure of alloys. The petroleum refinery process equipments such as furnaces, tubes, valves and pipelines that frequently operate at high temperature are usually more prone to CSCC [17]. Hydrogen sulfide (H_2S) also causes SCC damage to the austenitic stainless steel equipments throughout the refinery. In fluidized catalytic cracking unit (FCCU), fractionating tower, absorber column, stripper column and connected heat exchangers are also prone to H₂S cracking problem. This cracking is typically intergranular with little branching and in some cases it may contain transgranular portions [18]. Polythionic acid is formed during refinery shutdown as a result of interaction of sulfide scale with moisture and oxygen at ambient temperature. SCC that is caused by polythionic acid is intergranular in nature and occurs when changes in the alloy grain structure happen.

In four natural sour gas plants, Singh, V. [19] have identified a total of 13 SCC failures on austenitic stainless steel equipments. Surprisingly, all of the failures occurred on 304, 304L and 316L stainless steel types. The microstructure tests revealed CSCC, which was attributed to the presence of a significant amount of chlorides in the wet sour gas. Singh, found that failures occurred in the condensate stabilization and vapor compression sections of condensate recovery unit in the sour gas plant.

A constant load method has been used by Nishimura *et al.* [20] to investigate stress corrosion cracking of sensitized 304 stainless steel type in hydrochloric acid solution. They noticed that the fracture appearance for specimens sensitized at temperature < 1100 K was intergranular and/or a mixture of intergranular and transgranular mode. However, above this temperature, the fracture mode was only transgranular.

Kamaya *et al.* [21] investigated the initiation behavior of stress corrosion cracking for sensitized Type 304 stainless steel in high temperature water. They used a constant load SCC test method combined with crack observation technique. They clearly observed multiple cracks initiated on the surface of the specimen. They concluded that the change in the crack initiation time and the distribution of cracks were dependent on the applied stress.

Ayodogdu et al. [23] studied the susceptibility to intergranular corrosion and electrochemical reactivation behavior of AISI 316L type stainless steel. They found an increase in anodic current with increasing potassium thiocyanate content in the test solution, but for the reactivation current a more complex behavior was seen. It was attributed to the formation of metastable pits during the anodic scan of the test procedure.

Almubarak et al. [24] in their work; tensile stressed sensitized specimens were exposed to different solutions that usually exist in the petroleum refineries environments. Experimental work that included both metallographic and electrochemical has been carried out on sensitized austenitic stainless steel specimens. Stressed sensitized 304, 316 and 321 stainless steel have been selected and subjected to various environments that included polythionic acid, sour solution and chloride solution which have been prepared in the laboratory to simulate the petroleum refinery environment. Microstructure investigation revealed a more severe SCC in polythionc acid than in the sour and chloride solutions. Type 321 gave better resistance to SCC than 304 and 316 in the three solutions. It was concluded that acidity of solutions has a relatively minor influence in promoting cracking; however polythionic acid was found to be the primary causative agent. Most refinery equipment's are exposed to environments where polythionic acid could form due to the oxidation of sulfide scale during turnarounds. Also, the formation of acids, such as chloride acid and sour acid solution, is common when regeneration of a catalyst is performed on shutdowns [25]. These chemicals, frequently, exist in the current stream of most units in petroleum refineries. Austenitic stainless steels have been found to be susceptible to stress corrosion cracking (SCC) under conditions involving high stress, sensitization due to high temperature exposure, and the presence of those chemicals. Obviously, SCC is more severe in polythionc acid than in chloride and sour solutions [26]. On the other hand, cracking needed many days to occur in the chloride and sour solutions. This strongly suggests that the acidity of the solution has relatively minor effect in promoting cracking and that the polythionic acid is the primary contributory agent.

Types 304 and 316 of stainless steel are unsuitable for long term exposure in sulfide service above 400 °C because of sensitization which results in susceptibility to polythionic acid stress corrosion cracking. Type 304 stainless steel is also more susceptible to chloride stress corrosion cracking after exposure to elevated temperatures.

Passivity of sensitized 304 austenitic stainless steels is more difficult in polythionc acid than in chloride and sour solutions. A second anodic current peak was observed clearly in the polarization curves for the sensitized types 304 and 316 of stainless steels, indicating to the existing of a crack and/or a severe SCC [24]. The second anodic current peak was not observed in type 321 in the chloride solution which meant the absence SCC, and that was confirmed by the microstructure investigation. Another conclusion that Almubarak *et al* team added is, the highest second current peak indicated a more severe SCC occurring at a faster cracking time. Anodic polarization curves provided a rapid and efficient non-destructive testing method for showing passivity and to determining SCC for stainless steels exposed to the common solutions in refinery environment.

Obviously, sensitization accelerate SCC and cause cracking easily especially in polythionc acid. Type 321 is considered the best alloy to resist SCC in most solutions compared with other types. Type 321stainless steel is commonly used as weld overlay material for high temperature sulfide corrosion resistance. It is entirely satisfactory for lighter gage service such as heater tubes, exchanger bundles, and linings. However, NACE Standard [27] for the protection of austenitic stainless steels during shutdown of refinery equipment is highly recommended. Type 321 stainless steels have been very satisfactory in high temperature sulfide service.

References

[1]- Harold M. C. (2010), The History of Stainless Steel, ASM International, NY, USA.

[2]- Production process, 2019 https://www.worldstainless.org.

[3]- Katak, H. and Raj, B. (2002), Corrosion of Austenitic Stainless Steels Mechanism, Mitigation and Monitoring, Narosa Publishing House, New Delhi, India.

[4]- Brown B. F. (1981), Stress Corrosion Cracking Control Measures, NACE. Publication, 2nd. Edition, Texas, USA.

[5]- Chandler, K.A. (1985), Marine and Offshore Corrosion, Butterworth Co., London, UK.

[6]- Moller, G.E. (1977), "The successful Use of Austenitic Stainless Steel in Sea Water", *Society of Petroleum Engineers Journal*, Vol. 4, pp. 35-45.

[7]- Kopliku, A. and Mendez, C. (2010), "316 Stainless Steel Instrument Tubing In Marine Applications-Localized Corrosion Problems and Solutions", *Proceedings of NACE Corrosion Conference, Corrosion'10,* Houston, TX., Paper No. 10305.

[8]- Ahmad, S., Mehta M. L, Saraf, S. K., and Sarawat, P. (1985), "Electrochemical Studies of Stress Corrosion Cracking of Sensitized AISI 304 Stainless Steel in Polythionic Acids", Corrosion, vol. 41, No. 6, pp 363-368.

[9]- Ross, R.W. (2000), "New Technology Stainless Steels and Nickel Alloys for Marine Applications in The Year 2000 and Beyond", *Proceedings of Oceans 2000 MTS/IEEE Conference and Exhibition,* Vol. 3, pp. 1597-05.

[10]- Ritzenhoff, R. and Hahn, A. (2012), "Corrosion Resistance of High Nitrogen Steels", Corrosion Resistance, Chapter 3, Edited by Hong Shih, Publisher: InTech., Shanghai, China.

[11]- Welding Research Council, WRC Bulletin. No. 138, (1969) "Intergranular Corrosion of Chromium-Nickel Stainless Steels- Final Report", New York.

[12]- Rhodes, P.R. (2001), "Environment-Assisted Cracking of Corrosion-Resistant Alloys in Oil Production Environments: Review", *Corrosion*, Vol. 57, No. 11, pp.923-52.

[13]- Tervo, J.; Romu, J.; Hamalainen, E.; Hanninen, H. and Liimatainen, J.(1977), *Proceedings of the Fifth Inter. Conference on Advanced Particulate Materials and Processes*, West Palm Beach, Fl., pp. 317- 29.

[14]- Almubarak, A.; Abuhaimed, W. and Almazrouee, A., "Corrosion Behavior of the Stressed Sensitized Austenitic Stainless Steels of High Nitrogen Content in Seawater", International Journal of Electrochemistry, Volume April, 2013.

[15]- Simmons, J.W. (1996), "Overview: high-nitrogen alloying stainless steels", *Material Science & Engineering*, Vo. A207, pp.159-69.

[16]- Jargelius, R.F. (1998), "Application of the Pitting Resistance Equivalent Concept to Some Highly Alloyed Austenitic Stainless Steels", Corrosion, Volume 54, No. 02.

[17]- Rhodes, P. R. (2001), "Environmental Assisted Cracking of Corrosion Resistance Alloys in Oil and Gas Production Environment: A Review", Corrosion, vol. 57, No. 11, pp 923- 952.

[18]- Ghosh, D. (2007), "Wet H$_2$S Cracking Problem in Oil Refining Processes- Material Selection and Operation Control Issues", Tri-Service Corrosion Conference, Department of Defense, Washington DC, USA.

[19]- Singh, V. (2004), "Performance of Austenitic Stainless Steels in Wet Sour Gas-Part 2", Material Performance, vol. September, pp 46-50.

[20]- Nishimura, R.; Katim, I. and Maeda, Y. (2001), "Stress Corrosion Cracking of Sensitized Type 304 Stainless Steel in Hydrochloric Acid Solution- predicting Time- to failure and Effect of Sensitizing Temperature", Corrosion, vol. 57, No. 10, pp 853-862.

[21]- Kamaya, M. and Haruna, T. (2006), "Crack Initiation Model for Sensitized 304 Stainless Steel in High Temperature Water"; Corrosion Science, vol. 48; pp 2442-2456.

[22]- Aydogdu, G.H. and Aydinol, M.K. (2006), "Determination of Susceptibility to Intergranular Corrosion and Electrochemical Reactivation Behavior of AISI 316L type Stainless Steel"; Corrosion Science vol. 48; pp 3565-3583.

[24]- Almubarak, A.; Belkharchouche, M. and Hussain, A. (2010), "Stress Corrosion Cracking of Sensitized Austenitic Stainless Steels In Kuwait Petroleum Refineries", *Anti-Corrosion Methods and Materials Journal*, Vol. 57, No.2, pp. 58-64.

[25]- Mohamed, N.A.; Johnston, W. and Howard, C., (2008) "The Impact of Hydrocracker Reactor from Processing of High Sulphur and High Acid Crudes"; 12th Middle east Corrosion; Manama, Bahrain.

[26]- ASTM G 35-98 (2004), Standard practice for "Determining the Susceptibility of Stainless Steels and Related Nickel-Chromium-iron Alloys to Stress Corrosion Cracking in Polythionic Acids" Designation:, ASTM International, PA, USA

[27]- NACE Standard, RP0170-2004, (2004) Protection Austenitic Stainless Steel, NACE Publications, Texas, USA :

www.ingramcontent.com/pod-product-compliance
Lightning Source LLC
Chambersburg PA
CBHW051938210526
45473CB00006B/2293